PHANTOM

AEROGUIDE 13

McDonnell Phantom FG Mk 1/FGR Mk 2

INTRODUCTION

Any aircraft with a production run of over 5000 must rate as pretty successful; if it also happens to have been designed and built outside the Communist bloc, free from the influences of State control, then it must be rather special. The Phantom is special. It is one of those few postwar aircraft that have combined the virtues of good basic design with an adaptability that has made them effective in a variety of different roles.

The multi-role aircraft – for that is what the Phantom is – is a fashionable concept these days. With the cost of a modern fighter matching that of a prewar battleship (and only a handful of navies could afford those), it is obviously sensible to attempt to produce a design that will tackle two or more tasks instead of one. This in itself is not the major problem – indeed, multi-role aircraft were commonplace prewar, especially aboard aircraft carriers where stowage space was at a premium. Very much more difficult is drawing up a machine that will perform its diverse tasks without scaling down the nature of the tasks themselves. Not that any such design is anything other than a compromise (as all aircraft are): the successful machine is the one that offers the best compromise – the one that makes the fewest concessions in pursuit of the ideal.

It is sometimes the case that the successful multi-role aircraft is the one that was originally designed for just a single specific mission. One thinks of the Hunter, an air superiority fighter that made a very good ground attack machine. One also thinks of the Phantom in this context, since although its original specification was a bit open-ended it was nevertheless primarily an attack aircraft and it went on to become, amongst other things, a first-rate interceptor. The fact that it was exclusively a Navy design doubtless helped. Aircraft that have to stand up to what are nothing less than controlled crashes every time they land on board a carrier just have to be super-strong, and they therefore tend to last longer than their more fragile land-based counterparts; if they are really outstanding in their own right, like the Phantom, they may also cross the great divide that traditionally separates navies and air forces, and be accepted for service even by the world's

Left: Carrier approach for a Royal Navy Phantom FG Mk 1. The full range of the aircraft's high-lift devices – leading-edge slats, blown flaps and drooped ailerons – is displayed in this photograph, all helping to keep the machine airborne at the lowest possible speed. *Royal Navy*

Below: Despite the low-level tactical camouflage worn by this aircraft, No 43 Squadron's Phantoms have always been assigned air defence as their main duty.

Right: A 1985 photo of a Royal Air Force Phantom FGR.2, the famous phoenix marking of No 56 Squadron worn proudly on its nose.

greatest land-based air arm. And of course if the Marines buy it, it must be okay.

How and why the Phantom came to wear British roundels is related in the next chapter. The story is the usual tale of political cack-handedness involving attempts to cut costs (which didn't quite work out) and high-level decisions taken in furtherance of socio-economic pragmatism, but that is now history and the fact is that the Royal Navy and the Royal Air Force got themselves a rugged workhorse that has seen service for very many years and will continue to fly in RAF colours for a while yet. In Fleet Air Arm terms, the aircraft is inseparable from HMS *Ark Royal*, the last of the Royal Navy's fleet carriers and the only ship ever to take an FAA Phantom squadron to sea. It was also widely regarded as the last of the Royal Navy's fixed-wing fighters (added poignancy), but the Harrier concept eventually laid this supposition to rest. In RAF service, the Phantom was first flown primarily as a ground attack and reconnaissance fighter, replacing the Hunter, but with the arrival in the mid-1970s of the Anglo-French Jaguar to fulfil this mission the emphasis switched to air defence and the American aircraft began gradually to supplant the shorter-range Lightning.

Elsewhere the aircraft is combat-proven, having shown its prowess particularly in Israeli service and, more prominently, during the Vietnam War. It is also a factor in the long drawn out Iran-Iraq War, although its efficacy is presumably hampered by problems with spares and general back-up facilities. It has not actually been flown in anger by the British services, but, since hostilities over the Falkland Islands are not, at the time of writing, officially closed, the tasks of the Phantom squadron based there must at times be close enough to 'the real thing'. For the present, the FG.1s and FGR.2s based in the United Kingdom assume the major responsibility for the defence of the country's air space. Working in conjunction with Victor tankers, they range far out over the North Sea, the Atlantic and even the Arctic Ocean, investigating potential intruders and reminding a would-be enemy that his presence is always noted. Just routine maybe – but vital.

DESIGN & DEVELOPMENT

Back in 1954 the McDonnell Aircraft Company of St Louis, Missouri, proposed to the US Bureau of Aeronautics (BuAer) a project designated AH. The aircraft in question was a large, single-seat, twin-jet affair, armed with four 20mm cannon and showing a generous array of external stores stations. There didn't seem to be much chance that BuAer would jump at this idea, since their energies at the time were being directed towards the need for a fast CAP (Combat Air Patrol) machine which, because of the drastic shrinkage in the size of nuclear weapons, might also have a secondary attack capability: the primary armament would be long-range Sparrow missiles, which could be swapped for bombs and extra fuel as the occasion demanded, and all the ordnance would be fitted internally. The CAP study was aimed at producing a replacement for the F3H Demon general-purpose, all-weather fighter, and maybe the fact that this aircraft was a McDonnell product had something to do with BuAer looking at the AH, looking at their own paperwork, and then suggesting to McDonnell that perhaps the two projects were not so far apart. By dispensing with its J65 (Sapphire) engines and substituting the more powerful J79, getting rid of the cannon armament, hanging on some missiles, reworking the fire control systems and installing better radar equipment (and a second crewman to cope with it), the AH would be just the ticket.

A year earlier McDonnell had suffered a serious blow when their submission for the Navy's next air superiority fighter lost out to the Vought F8U, threatening to bring to an end a successful line of carrier-based aircraft designs. It was no great surprise, therefore, when BuAer's suggestion was seized upon. In late 1954 two AH prototypes (YAH) were ordered, and the various design modifications were incorporated into these through 1955; other refinements were introduced, including the now famous wing-tip dihedral and stabiliser anhedral. The project was redesignated F4H, a full-size mock-up started to be put together, and things really began to hum.

By mid-1956 the design had been finalised, and work started on the XF4H prototype. By early 1958 the aircraft had been rolled out, and the first flight took place on 27 May that year. BuAer contracted for 23 pre-production aircraft for the US Navy, and more refinements were progressively worked into the design as these machines began to take shape, including particularly the fitting of even better radar equipment. A neat compromise reached earlier concerned the missile armament: external with the AH and internal in BuAer's original specification, the final layout saw the weapons semi-buried below the fuselage.

Carrier compatability trials were conducted on board USS *Independence* during 1960, and later that year, with the aircraft's future now assured following its official selection as standard Navy equipment, the first production machines (F4H-1) were completed, against an initial order for 24. Intensive training with the new type followed in 1961, and the aircraft was accepted into operational service after only a few months; by late 1963 it was in regular use not only by the US Navy but also by the Marine Corps and the Air Force. The name came about just before the XF4H's maiden flight. Old hands had heard it before, since an earlier Phantom, the McDonnell FD/FH, had seen limited service back in the late 1940s and had won itself a permanent place in the history books for being the first US jet aircraft to land aboard a carrier.

The fact that the Royal Air Force currently operates Phantoms can, as with its Buccaneer strike bombers, partly be explained by political expediency, but the beginnings of British interest in the aircraft hark back to

Far left top: The first prototype F4H-1 Phantom, which made its maiden flight on 27 May 1958. Production machines would show some changes, including a broader nose to carry the main radar, a raised cockpit to improve the crewmen's view, and redesigned main intakes. *McDonnell Douglas*

Far left bottom: XT595 was the first of two YF-4K prototypes, and is seen here making its first flight, 27 June 1966. The forward missile bays are equipped with dummy Sparrows. *McDonnell Douglas*

Above: The third production Phantom for the Royal Navy, XT857. No serial has been applied to the fuselage, but the constructor's number, 2097, appears at the top of the fin. *McDonnell Douglas*

Below: The first YF-4M, XT852, on the ramp at the manufacturer's plant at St Louis. The aircraft is in grey/green camouflage and is fitted here with an F-4K-type nosewheel door, complete with carrier approach lights. Inboard pylons are fitted and auxiliary engine doors open. *McDonnell Douglas*

1960 and the simultaneous (though separate) requirements for, first, an aircraft to replace the Royal Navy's Sea Vixen all-weather fighter and, second, a successor to the RAF's ground-attack/reconnaissance Hunter. In order to save on development costs (so the theory went), it was considered that a single design might be employed as a basis to fulfil both requirements, and eyes fell upon the Hawker P.1154, an intended supersonic follow-up to the V/STOL P.1127 then flying. Development began in earnest in 1963, and the following year the construction of two prototypes, a dual-seater for the Royal Navy and a single-seater for the RAF, was begun. But already some frowns were appearing across Navy brows. For one thing, a big shiny Phantom prototype, having recently amazed all and sundry at the Paris Air Show, just happened to land at the Royal Naval Air Station at Yeovilton in Somerset, on its way back to the United States. It looked impressive – and available. For another, McDonnell kept pushing around official reports with come-hither titles like *F-4 for the Royal Navy* which showed how absolutely terrific the Phantom would be in Fleet Air Arm colours. Then again, the price of re-equipping the FAA with the space-saving P.1154 could mean the premature demise of the Royal Navy's fleet carriers. And the P.1154 would mean a wait of six years before the Navy got its new aircraft – far too long. Enthusiasm evaporated, and by late 1963 the Phantom was being backed to the hilt.

The P.1154 held on for several more months, but shortly after the new Labour administration came into office in November 1964 the political axeman was unleashed against homespun military aircraft projects once more: out went the supersonic V/STOL, together with the Hawker Siddeley HS.681 STOL transport which would support it; the substitute for the P.1154 would be the Phantom, and thus by a strange irony the Royal Navy and the Royal Air Force would get a common airframe after all.

The Phantom was designed essentially for operation from aircraft carriers, but a number of technical modifications had to be worked out in order to make it suitable for use with the British fleet. A problem quickly appreciated was that, as it stood, the F-4 simply would not fit existing carrier lifts, so the nose radome was altered to enable it to hinge sideways, but that little difficulty paled in comparison with a more fundamental question: how could the Phantom be made to operate from the short flight decks of the RN carriers? Flying the aircraft even from the US *Essex* class ships (roughly comparable for these purposes to the Royal Navy's largest units, *Eagle* and *Ark Royal*) was at best marginal, but the F-4 was supposed to equip the far smaller *Hermes* as well. Some pretty drastic modifications would be needed, both to the ships and to the aircraft.

The Phantom was a large and very powerful aircraft, and it had an extremely high landing speed: it was therefore imperative to see if the latter could be cut, to take account of the reduced landing area and less substantial arrester gear. It was just as vital to try and compensate for the British carriers' less powerful catapult equipment, whilst in both tasks the slightly inferior speeds of the RN vessels gave less wind-over-deck (WOD), which in itself had the effect of requiring lengthier distances for both take-off and landing.

The take-off problem was solved partly by introducing an extra-extensible nosewheel leg, which gave the aircraft's wings an increased angle of attack (the famous Harrier-type 'ski jump' has a related function), whilst low-speed handling was decidedly improved by fitting larger flaps, drooped ailerons and fixed leading-edge slats on the tailplane (features also adopted by the US Navy's F-4B successor, the F-4J). Other changes, notably more space for fuel, which had been introduced with the USAF's F-4E upgrade programme and again incorporated into the F-4J, appeared in Royal Navy Phantoms, giving an increased all-up weight and calling for stronger landing gear.

Phantoms ordered for the Royal Air Force (F-4M) did not receive the carrier-dictated modifications worked into the Fleet Air Arm's machines and retained the standard nose gear and non-slotted stabilisers. However, common to both versions was a far-reaching change, and one exclusive to British F-4s – the installation of Rolls-Royce Spey engines. The decision to use Speys can be partly traced to the desire to placate those opposed to the purchase of foreign air combat equipment and to ensure that at least some of the work involved in getting the F-4 into British service would be done in manufacturing plants at home. An important consideration, however, was the undeniable superiority promised by the Rolls-Royce engine in certain important respects, for example greater economy of fuel consumption (hence better range) and greater available thrust (giving far better acceleration, especially at low altitudes).

Unfortunately for the Government's accountants, and to the frustration of the admirals, the marriage of the Spey to the Phantom airframe was not an entirely happy one, and some major redesign headaches caused so much delay and extra cost that the two main reasons given by the Government for plumping for the Phantom in the first place – availability and cheapness – threatened to look a bit silly. The entire engine bays had to be widened and deepened to accommodate the new powerplants, and the thrustline had to be canted downwards an extra degree or so, improving the aircraft's take-off characteristics even more but requiring some expensive recontouring of the after fuselage; even so, the incredible heat thrown out by the Speys' afterburners meant that more titanium insulation had to be fitted beneath the tail. The air-thirsty Speys

Left: XT852 in October 1967, some months after it began its flying trials. The forward missile bays are here fitted with EROS, making the aircraft 'collision avoidance equipped' as the legend states. This aircraft was one of six pre-production Spey-engined Phantoms used for pre-delivery test purposes. *McDonnell Douglas*

Right: Phantom rear fuselages under assembly by the British Aircraft Corporation at Preston. The production of RN and RAF Phantoms involved a fair amount of sub-contract work for British companies, the most valuable element by far being the Rolls-Royce engines. Other major items included the Ferranti inertial navigation/attack system (INAS), the Martin-Baker ejection seats and the outer wing panels, the last being manufactured by Shorts at Belfast. *British Aerospace*

Below: Three Phantom FG.1s were employed in catapult and arrested landing trials on board the carrier *Eagle*. Here one of them, XT865, shows its extra-extensible nose gear. *Royal Navy*

also needed bigger main intakes, whilst auxiliary intake flaps had to be installed further back along the fuselage in order to give the aircraft a satisfactory taxying performance.

An order for the F-4K was announced on 1 July 1964. The go-ahead was at first limited to a couple of prototypes, but the general feeling was that the eventual total bought would be about 140 aircraft; within a year the RAF procurement of F-4Ms, involving 150 Phantoms, had also been announced. Two prototypes of each version (YF-4K and YF-4M) were contracted for, and these would be entirely US-built, apart from the Spey engines.

The maiden flight of the YF-4K took place on 27 June 1966 and that of the YF-4M on 17 February the following year, by which time the other YF-4K and a pre-production F-4K had joined the trials programme. A further F-4K and the second YF-4M were added during 1967, and the six aircraft were put through exhaustive flight tests wherein the problems afflicting the Spey were gradually resolved and some of the other snags (including the incompatibility of the US AWG-10 fire control radar with the British nav/attack system of the YF-4M) were sorted out.

Meanwhile British Government defence strategy took another twist, which began to affect the whole Phantom deal: spending economies were sought. The RAF order was cut back to 116 production aircraft and the Fleet Air Arm's requirement was savagely pruned. Owing mainly to a complete reorientation of naval policy, which included the cancellation of the new fleet carrier CVA-01 and told of the phasing out of all the old ones by the mid-1970s, the Royal Navy's Phantom allocation was slashed to one-third of the original estimate. Further economies came along later, and the Navy would actually receive just 28 machines, enough for only one full front-line squadron.

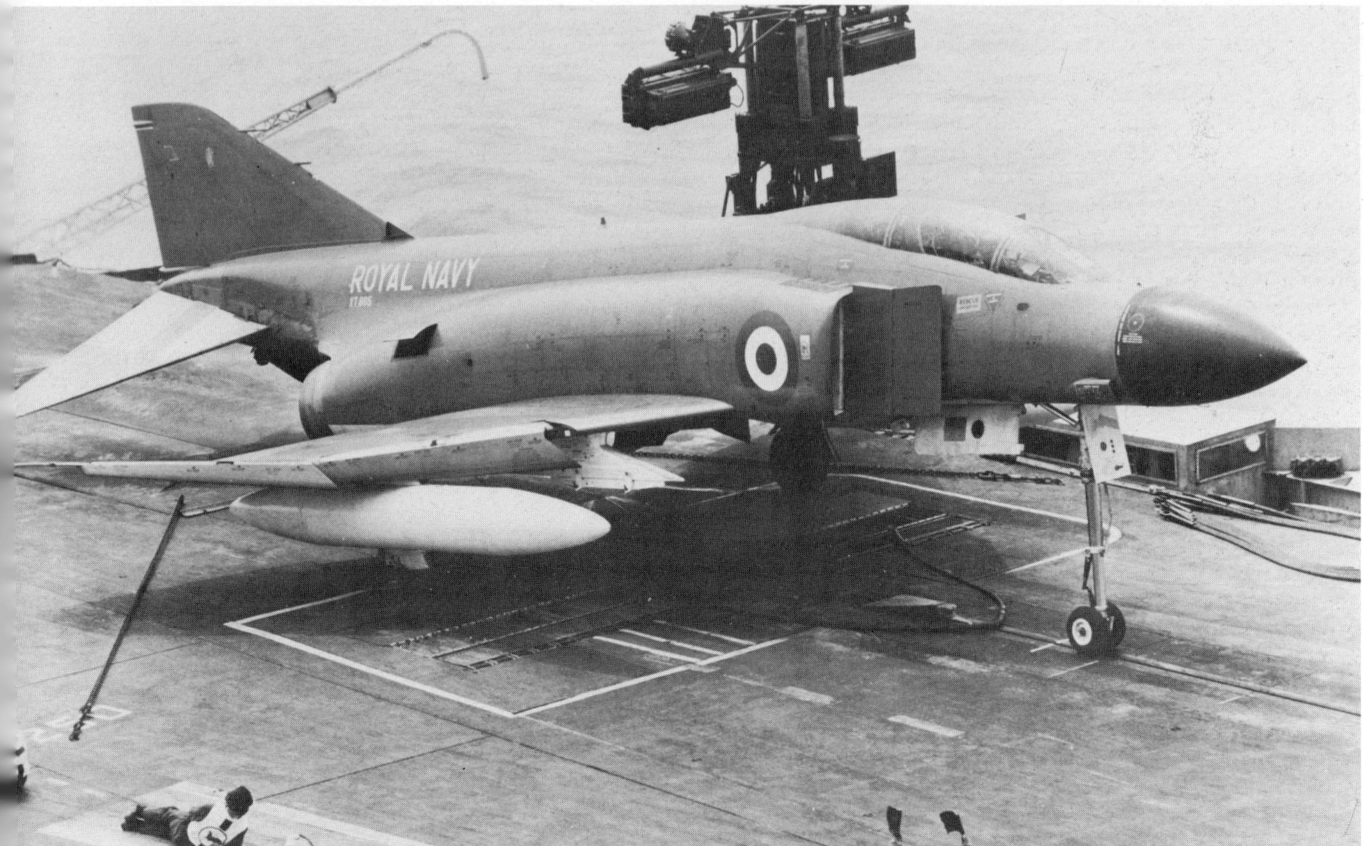

STRUCTURE

Apart from the changes involved in installing the Rolls-Royce Spey turbofans, already outlined, and the local modifications to the undercarriage necessary for Royal Navy FG.1s, the structural design of UK-operated Phantoms is little different from that of the US Navy F-4J, which was coming off the St Louis lines at the same time.

The forward fuselage houses the Westinghouse AWG-11/12 (modified AWG-10) radar dish and its associated equipment, the nose undercarriage gear and the two-place cockpit. The AWG-11 (employed in Royal Navy aircraft) and -12 (RAF) are impressive in both size and capabilities. The system is a true multi-mode one, able to conduct air-to-air search and tracking, air-to-ground 'look down' and ranging, and mapping. It also has a continuous-wave (CW) illumination mode to enable the Phantom's Sparrow or Sky Flash missiles to home on a target in front of the aircraft.

The centre fuselage houses the two Spey powerplants, fed from fuel tanks above the engine bays, in front of them (immediately behind the navigator's position) and within the forward section of the wing. The total gallonage is about 1990; put another way, this comes to something of the order of 13,250lb, or almost half the aircraft's empty weight. The wing itself is of tapered delta shape, with 45 degrees of sweep at the leading edge. Control surfaces consist of downward-motion ailerons along the trailing edge complemented by upward-motion paired spoilers forward of the aileron hinge line. The ailerons also function as limited lift devices, depressing in conjunction with the main flaps inboard. Leading-edge slats, operating in concert with the flaps, are distributed two per wing, along the outer edge of the centre section, and along the front of the wing-tip panels. The latter are one of the distinctive Phantom features, angled with 12 degrees of dihedral to maximise the aircraft's lateral stability. The other Phantom hallmark is the sharp anhedral (31 degrees) of the tailplane. The two stabilisers operate in unison.

The undercarriage is arranged in the traditional 'tricycle' fashion and is immensely strong. The great breadth of the wing allows ample space in front of the main wheel bays for stores pylons, and indeed for a pair of wing-mounted air brakes. An arrester hook, standard equipment for high-performance naval aircraft but also used with land-based runway arrester systems, is located along the keel of the rear fuselage section.

Below left: Forward fuselage of a Phantom FGR.2, showing the two-place cockpit and starboard intake duct.
Right: Port aspect of the forward fuselage, Phantom FG.1. The red ejection seat warning triangles are considerably larger on Phantoms than on other RAF aircraft.
Right below: Black-painted radomes have been abandoned with the adoption of greys as the overall scheme.
Right bottom: Nose ram air cooling inlet (left) and pilot's retractable foot step (right). The function of the yellow flashes on the intake is classified, but they are possibly indicative of the degree of upgrade the aircraft has received.

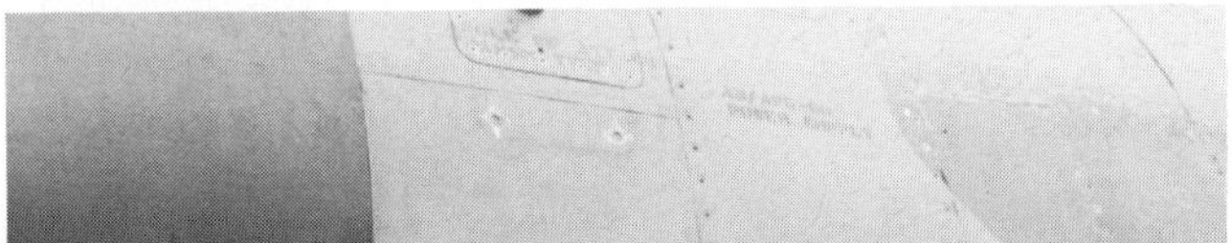

This page top: The ultra-smooth canopy profile, the lower photo showing the navigator's rear-view mirror fitted to some Phantoms.
Opposite page top left: A recent introduction on some aircraft is a powerful periscope, fitted between the hinged canopies on the port side.
Opposite page top right: An in-flight refuelling probe is installed along the starboard fuselage, beneath the navigator's canopy.
Below: Modern high-speed aircraft need some system to vary the amount of air reaching their engines, so as to extract the maximum performance across the wide range of altitudes at which the aircraft operates. As these four photos show, the Phantom has a perforated, variable-angle ramp immediately in front of each intake, altering the volume of air ingested. Forward of this is a broad, fixed splitter plate, which prevents turbulent boundary layer air skimming along the fuselage from entering the intake.

RESCUE
OPPOSITE SIDE

XT863

4
LEU R41

Opposite page: Two Phantom FG.1s from No 111 Squadron, showing the upper auxiliary air doors towards the back of the intake duct. The exhaust port immediately forward of the door serves as an in-flight reference point; the J79-engined F-4J (UK) Phantoms flown by No 74 Squadron do not have this feature, so an adhesive sticker is added to serve the same purpose. Both photographs were taken in the summer of 1985, when XT863 (top) was still wearing its special *Concourse d'Elégance* scheme adopted for the 25th Anniversary 'Phantom Meet' at the 1983 Greenham Common Air Tattoo. XT864 (bottom) has a replacement starboard outer wing panel and is shown in profile on the back cover of this book; it also appears on page 33, in its earlier Royal Navy colours.
Left above: Contrasting wing tip decor on two No 43 Squadron Phantoms, 1985.
Left below: Wing fold hinge.
Below: View over the port wing of an FG.1, showing the flaps and drooped ailerons.

This spread: Eight photos showing various angles on the Phantom's tail; the FG Mk 1s built for the Royal Navy can readily be identified by their slotted stabilisers, which gave greater control at the high angles of attack associated with carrier take-offs and landings. The downward angle of the stabilisers can be explained by the need to ensure good longitudinal stability whatever the wing's angle of attack, at the same time keeping them clear of both the wing wash and the jet efflux. The photos also show the retrofitted fairing on top of the fin, which houses passive radar warning receiver (RWR) equipment and is unique to British Phantoms, and the broad antenna beneath (another retrofit), part of the aircraft's ILS localiser/glide path system. The titanium heat shielding around the lower rear fuselage can be seen, as can the main fuel vent, located above the parachute brake housing at the base of the fin trailing edge. Along the leading edge of the vertical fin are the pitot tube (top), the artificial feel sensor for stabiliser trim, and the upper anti-collision beacon (which on most other types of aircraft is located on top of the fuselage). The arrester hook is hydraulically powered and was specially strengthened for British Phantoms, both versions of which are equipped with it.

J

E

NO PUSH

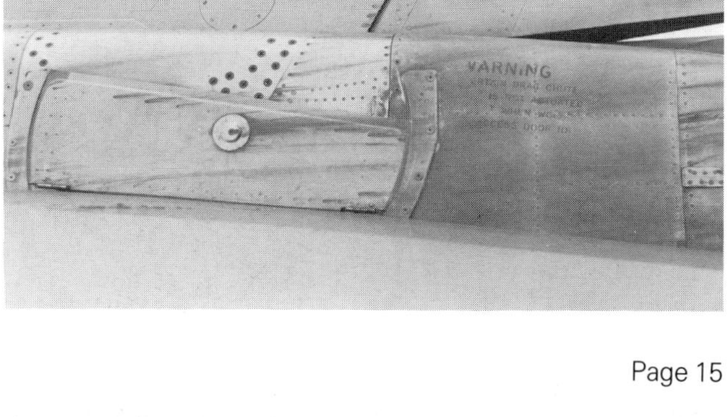
WARNING

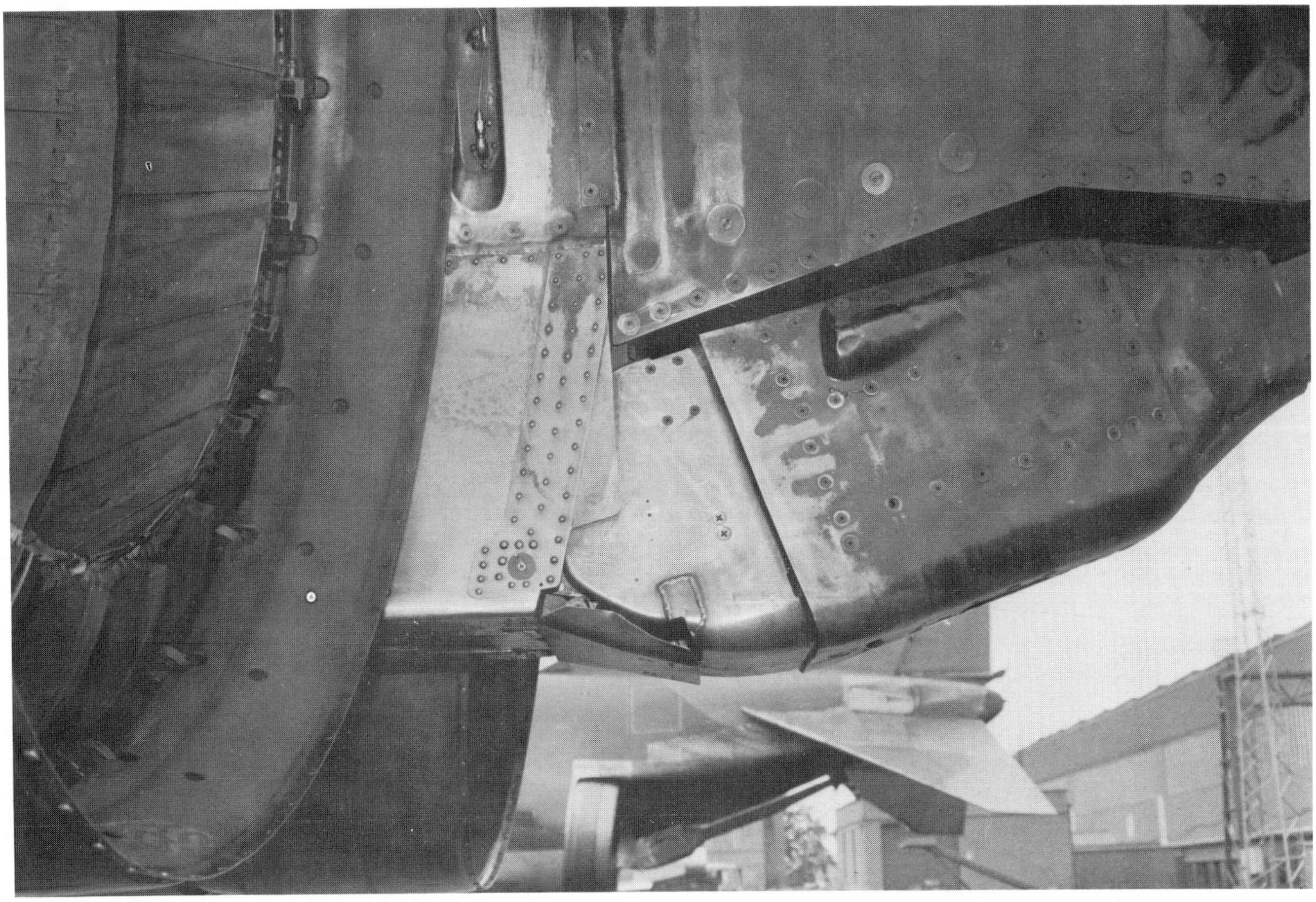

Above: The arrester hook attachment point, located between the jetpipes. The tailhook is often to be seen lowered to the ground whilst Phantoms are being re-armed.
Below left: The two Rolls-Royce Speys that power the Phantom FG.1 and FGR.2 are Mk 201 series engines, the Mk 101 series being the non-afterburning versions that were developed for use in the BAC.111 and Trident civil airliners and the Buccaneer strike bomber. It is a bypass turbofan: not all the air gulped through the main intakes is burnt in the core, some being mixed with the main exhaust stream after combustion to give greater propulsive efficiency, and some being bled off to provide air for boundary layer blowing. The Spey combines enormous thrust with notable fuel economy.

Below right: Close-up view of the fuel vent fitted at the extreme tail; other vents are fitted along the wing trailing edges.
Opposite page top: One of the 28 Phantom FG.1s that served with the Fleet Air Arm takes off from Lambert Field, St Louis. The three external fuel tanks suggest that the aircraft may be setting off on its transatlantic delivery flight. *McDonnell Douglas*
Opposite page bottom: Afterburners blazing, a Phantom FG.1 roars off one of *Ark Royal*'s catapults. The extra height given to the nose of the aircraft by means of its modified nose-gear is clearly evident. The Omega insignia adopted by 892 Naval Air Squadron did not, in the event, signify the last fixed-wing unit operated at sea by the FAA. *McDonnell Douglas*

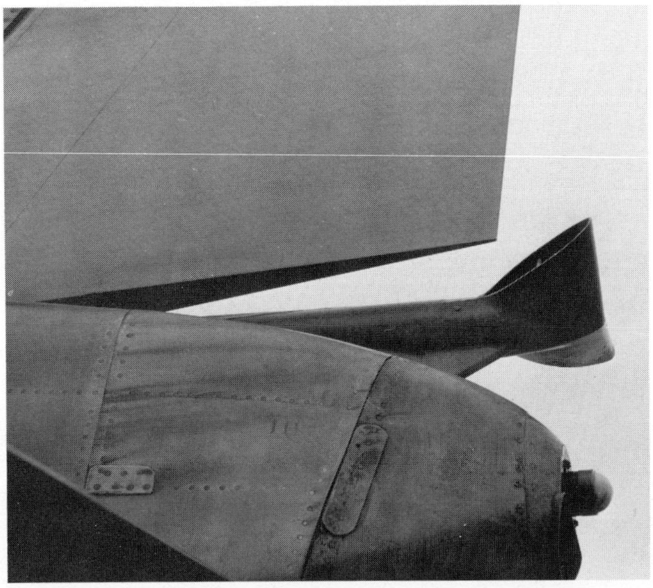

ROYAL NAVY
MCDONNELL

010
R
ROYAL NAVY

Above: As of late 1985, Phantoms finished in Dark Sea Grey/Dark Green disruptive camouflage were still to be found on the flight lines: this example, FG Mk 1 XV586/'J', is wearing the insignia of No 43 Squadron. Wing tips are frequently folded up on parked aircraft, in the interests of saving space.
Above right: The other Phantom FG.1 unit, No 111 Squadron, also retained tactical camouflage (on dwindling numbers of aircraft) well into the 1980s. Here XV592/'L', on standby and thus fully armed, is seen in July 1985.
Above far right: XT870/'S' shows the light grey low-visibility scheme currently in vogue for UK air defence aircraft, although the radome is, unusually, painted Medium Sea Grey. The changeover to pale colours has not, in the case of 'Treble-One's Phantoms, been accompanied by any permanent scaling down of the squadron markings.
Right: Phantom FGR.2 XV428/'I' undergoes routine maintenance in one of No 56 Squadron's Hardened Aircraft Shelters (HAS) at RAF Wattisham, August 1985. Notice the two styles of intake guards and, in the foreground, the earthing cable.
Far right: No 43 Squadron line-up, July 1985, showing to advantage the splendid 'Fighting Cock' unit emblem and the variety of metallic hues across the FG.1's slotted stabiliser. With the switch to protective shelters, sights such as this are gradually becoming a thing of the past.

X
XV590

RESCUE

B
5

Opposite page: The zenith of Royal Navy air power: HMS *Ark Royal* at sea with her Phantoms. When the ship decommissioned for the last time, her aircraft went to the RAF. *Royal Navy*
Below: Phantom FG.1 of 767 Naval Air Squadron, 1967. Finish is Extra Dark Sea Grey and white, with white lettering, fuselage serial and tail code. Squadron insignia is yellow with black detail.

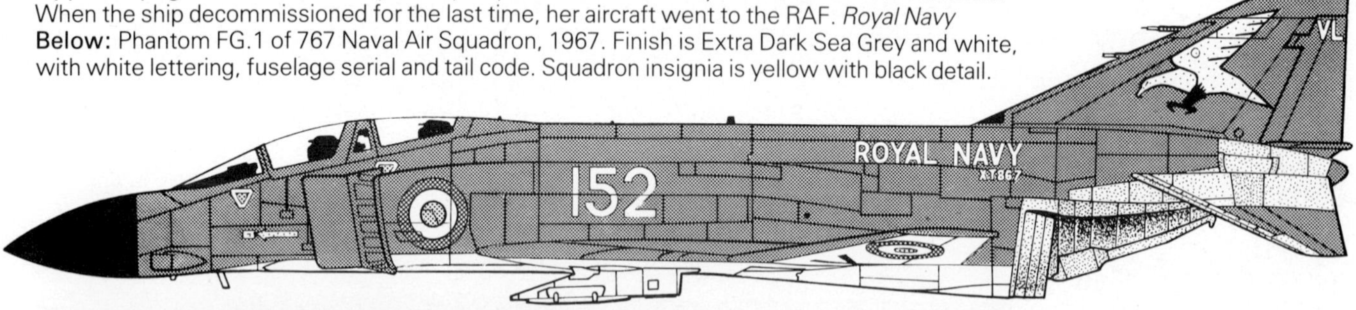

Left: Phantom FGR.2 XV418/'C', in the markings of No 41 Squadron and equipped for the recce role, 1972. The camouflage finish is the standard tactical Dark Sea Grey and Dark Green, with Light Aircraft Grey undersurfaces. *RAF Official*
Below: A No 17 Squadron FGR.2 in similar finish, with the unit's distinctive black and white marking on the intake.

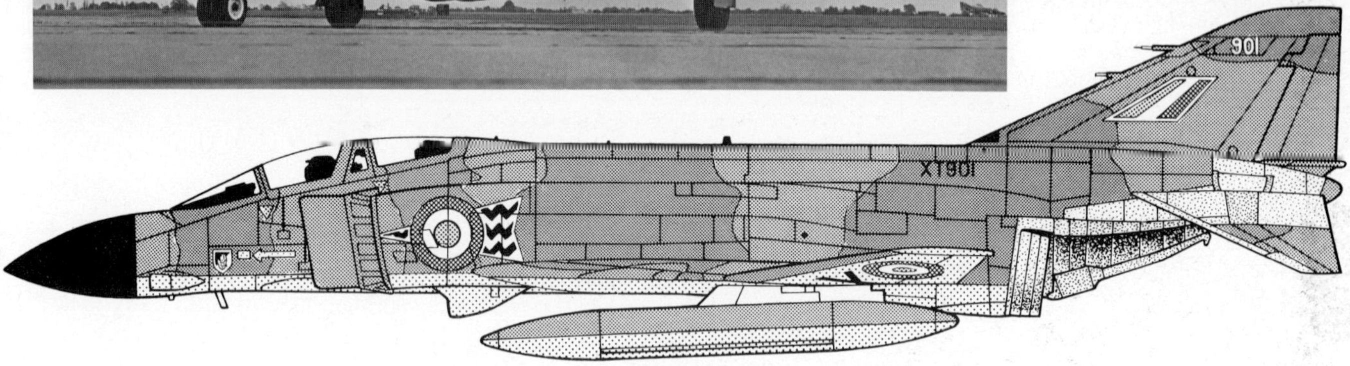

Left: FG Mk 1 XT864/'007' at St Athan just prior to transfer to the RAF; the Navy sign-writer has been at it! The red, white and blue nose flash was originally added to mark HM The Queen's Silver Jubilee, although the '77' logo was later replaced by the 892 NAS unit badge, as shown. From 1972, support and maintenance facilities for Navy Phantoms was provided by the RAF at Leuchars. *John Hale*

Below: A No 2 Squadron Phantom FGR Mk 2 with 'toned-down' roundels, a modification carried out in the mid-1970s. The unit markings on the nose are black and red, and the fin code letter is black on a red background.

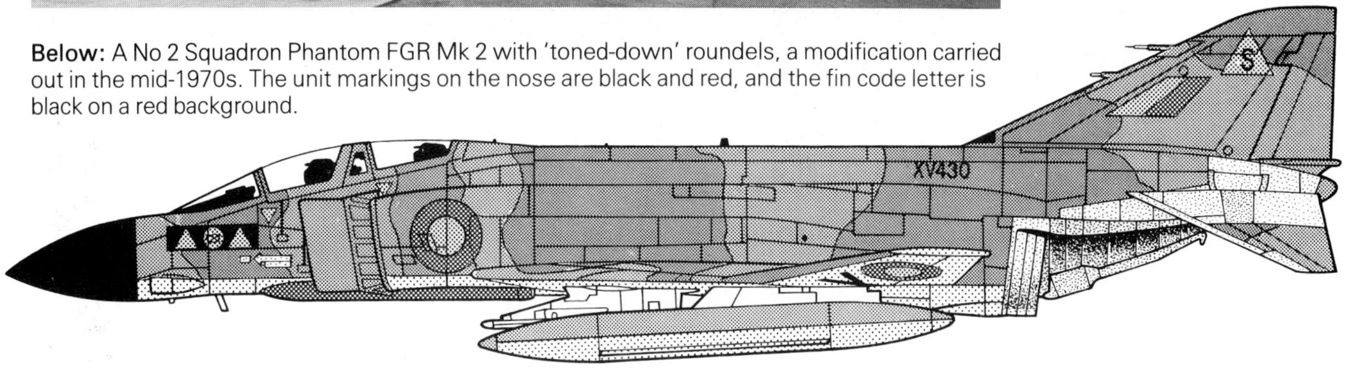

McDONNELL PHANTOM FGR Mk 2, No 56(F) SQUADRON, RAF WATTISHAM, AUGUST 1985

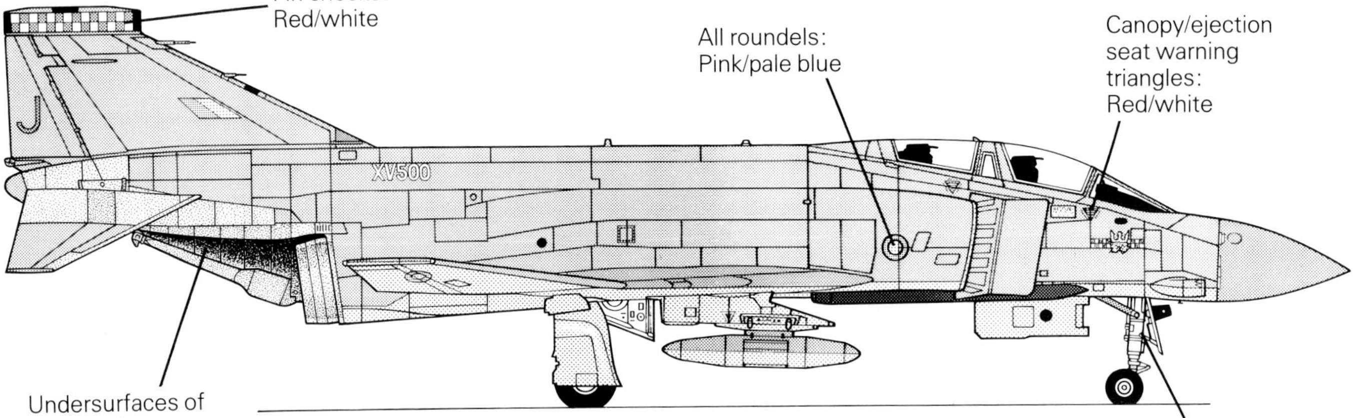

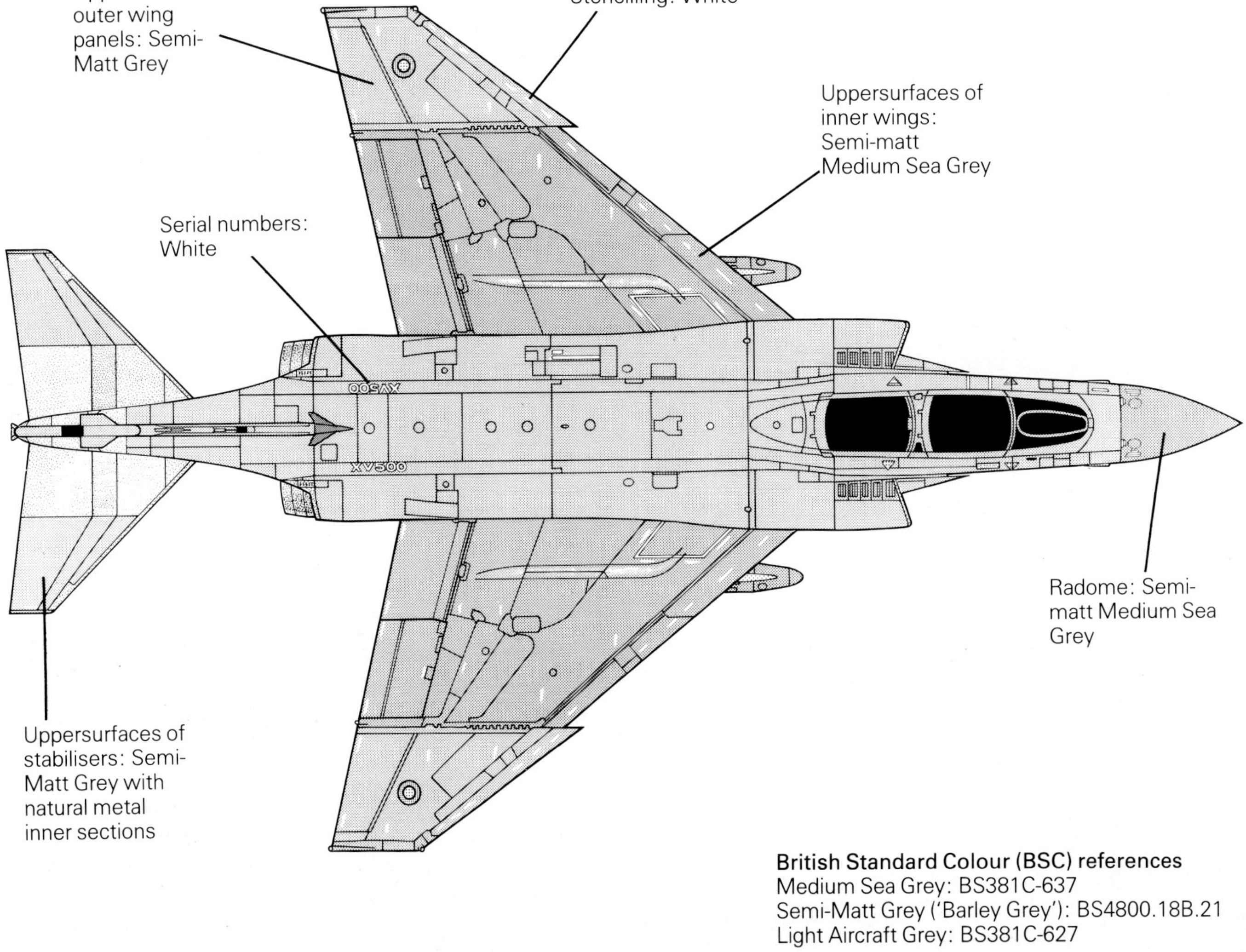

British Standard Colour (BSC) references
Medium Sea Grey: BS381C-637
Semi-Matt Grey ('Barley Grey'): BS4800.18B.21
Light Aircraft Grey: BS381C-627

1:96 scale

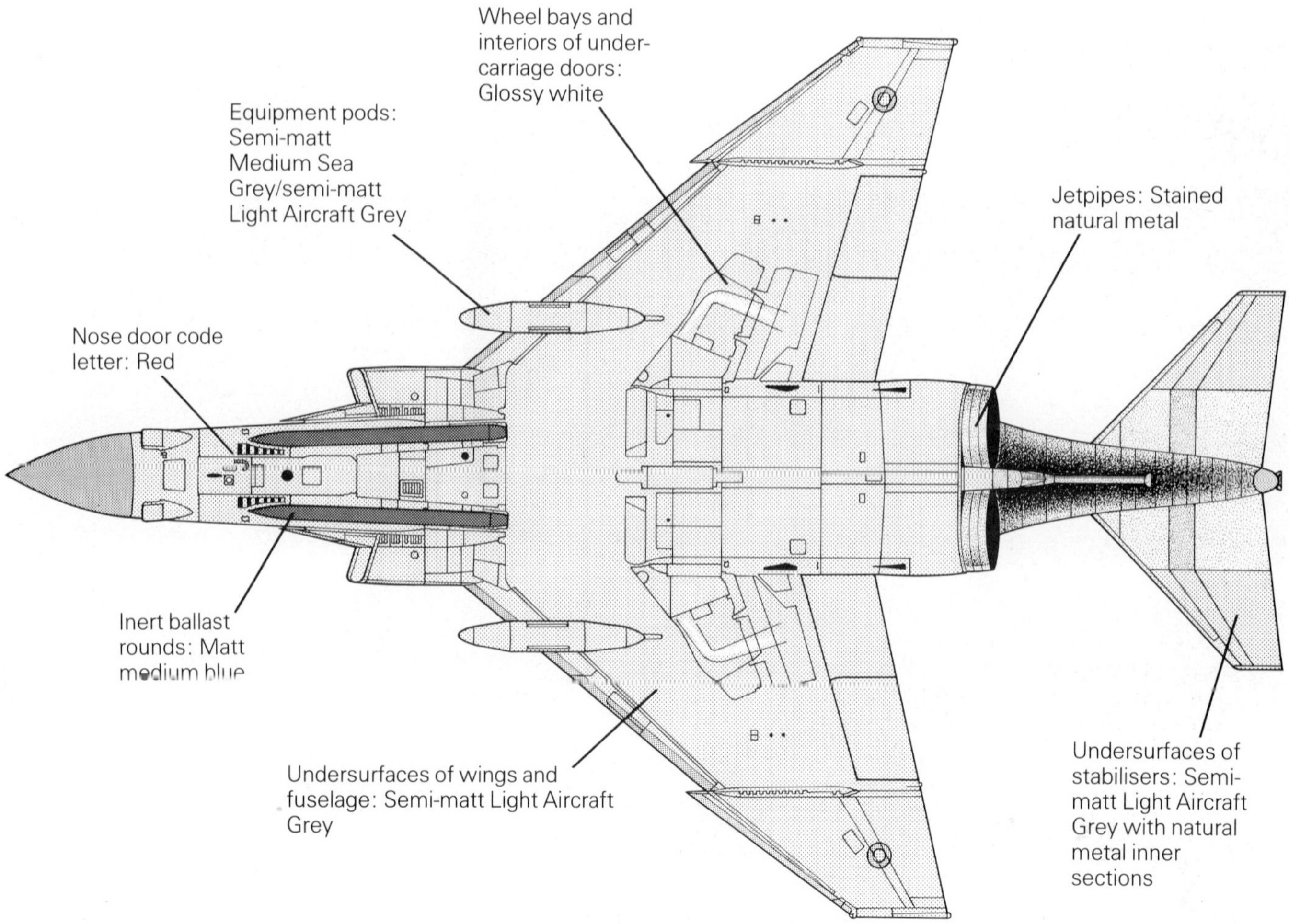

For further information on Phantom colour schemes and markings see Modeldecal sets 2, 23, 24, 27, 30, 37, 44, 55, 64, 65, 67, 71, 72, 73 and 76

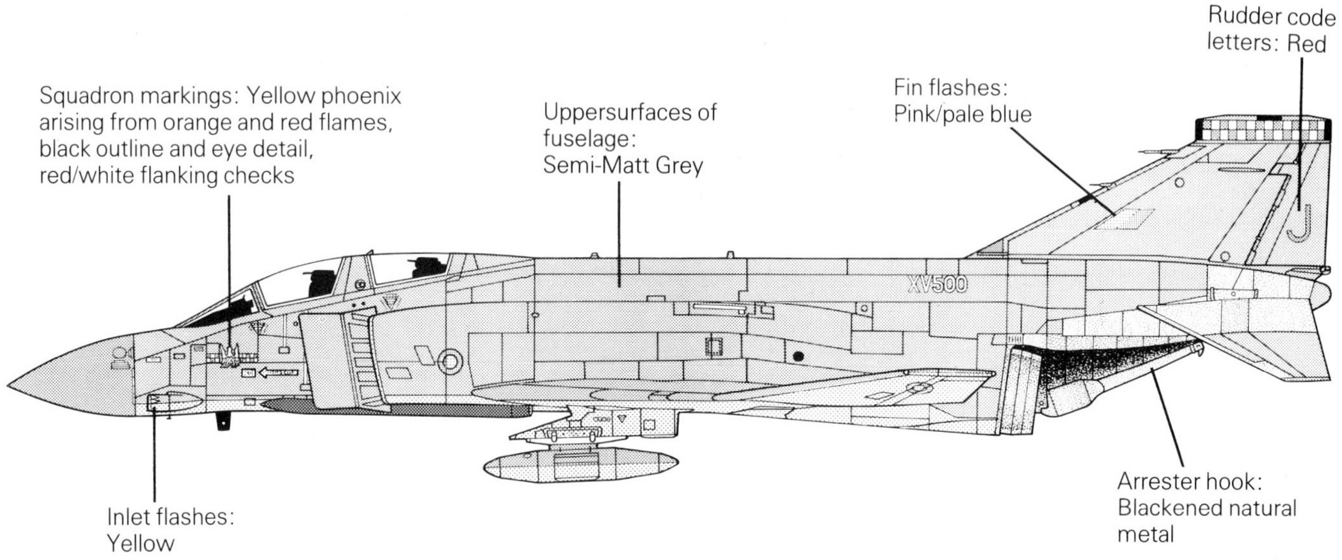

Below: Phantom FG.1 XT859, No 111 Squadron, shows the current Phantom scheme of Semi-Matt Grey ('Barley Grey'), Medium Sea Grey and Light Aircraft Grey, with muted squadron markings in grey on the tail. This particular aircraft, serving with the Fleet Air Arm, won the 1969 *Daily Mail* Transatlantic Air Race with a record west–east crossing.

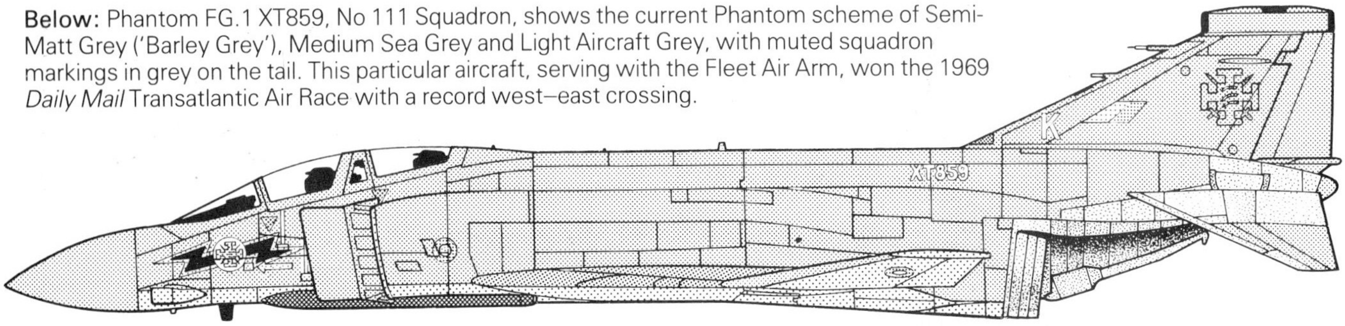

Left: Up on jacks for maintenance at Wattisham is FGR.2 XV492/'U' (No 56 Squadron), one of the last FGR.2s to be delivered to the RAF.

Below: XV480/'X' of No 92 Squadron, RAF Germany, 1984, shows a Medium Sea Grey radome but its finish is otherwise standard. The fin-top checks are red and yellow, whilst the cobra emblem on the fin is red and yellow with black detailing. No 92 is one of two Phantom units based in the Federal Republic.

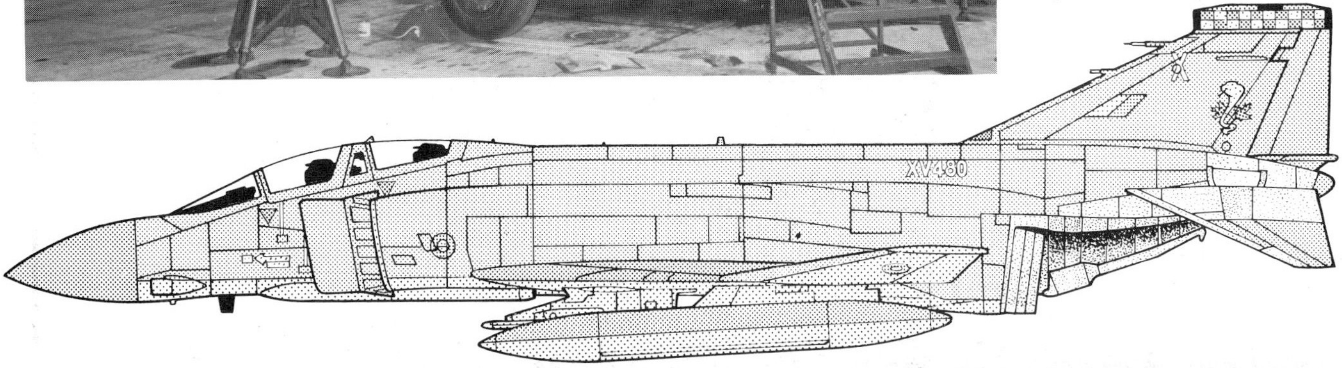

Right: Tiger ready to spring. Following the Falklands War of 1982, and the subsequent transfer of Phantoms to the South Atlantic, a batch of fifteen second-hand F-4Js was purchased from the United States to maintain UK air defence strength levels. The aircraft, designated F-4J (UK)s, were completely refurbished and taken on charge by No 74 Squadron, which was re-formed for the purpose in 1984. The aircraft differ considerably from the FG.1s and FGR.2s, not least in that they are fitted with J79 engines; spares and maintenance thus tend to be relatively expensive in comparison with the rest of the Phantom fleet. The aircraft shown, ZE363, is on 'Q' at Wattisham.